LIVING WITH LOGS / BRITISH COLUMBIA'S LOG BUILDINGS AND RAIL FENCES

LIVING WITH LOGS
BRITISH COLUMBIA'S LOG BUILDINGS AND RAIL FENCES

TEXT AND PHOTOGRAPHS BY
DONOVAN CLEMSON

ILLUSTRATIONS BY
SUSAN IM BAUMGARTEN

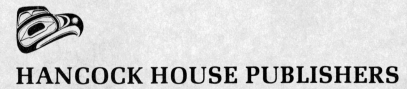

HANCOCK HOUSE PUBLISHERS

I.S.B.N. 0-919654-10-x

Copyright © 1974 Donovan Clemson
and Susan Im Baumgarten

All rights reserved. No part of this
publication may be reproduced,
stored in a retrieval system, or
transmitted, in any form or by
any means, electronic, mechanical,
photocopying, recording, or otherwise, without the prior written
permission of Hancock House
Publishers.

The kind cooperation of both the Saanich Pioneers Society and the Saanich Historical Artifacts Society is acknowledged for use of their photo and artifact collection.

SPECIAL NOTE TO THE READER

Do you possess old photographs, artifacts or old household items of the past? If they are laying wastefully in a trunk etc., please con- sider donating them to your local museum, historical society or to the Provincial Museum or Archives. The publisher became aware of the need for more of this material to be on display and available to the public when he undertook recently to prepare a series of books relating to our past. So much material is just not available and the regional and provincial museums ask that more people save these old photographs and items and donate them to the museums.

Design by McConnell Design

Typesetting by White Computer Typesetting
Vancouver, B.C.

PRINTED IN CANADA

Published by:
HANCOCK HOUSE PUBLISHERS
3215 Island View Road
Saanichton, B.C., Canada

CONTENTS

Introduction	6
The Primitive Touch	14
Relics of Pioneer Days	22
The Native Villages	36
Fences and Corrals	48
The Scandinavian Way	60
Living with Logs	72
Mainly Barns	84

INTRODUCTION

"The heathen in his blindness bows down to wood and stone" — so runs the line of the old hymn. If that is true, call me heathen, for I value those two materials above all others. This book deals with wood; wood in the rough, that never reached the sawmill or planer; wood scored with the axe and carefully dressed with the broadaxe; wood in the round, rough with bark or with the cheek-smoothness of logs peeled when the sap was rising; logs and rails, the traditional stuff of the pioneers.

With whole tree trunks, roughly dressed, the fur traders built their forts, the missionaries their missions, the settlers their homes, barns, sheds, root-houses, fences, gates and corrals. The interior native Indians, when they abandoned their underground winter houses, built villages of log houses of such sturdy construction that with proper maintenance they could all have been standing today.

For a hundred years the building of log houses flourished throughout interior British Columbia, and a few of the buildings existing today can look back more than a century to the day when their bottom logs were skidded out of the bush. Many more would still be standing if their owners had been interested enough to patch the roof or repair the foundations, for neglect of these, however sturdy the structure, brings decay and inevitable ruin. Many early cabins, built without foundations, slowly settled into the ground as the bottom log settled and rotted, lowering the doorway to an awkward height in the process. But the cabins were not always intended for permanent use. The backwoods settler built his cabin in a hurry, intending it as temporary shelter until he could get around to building a real home which would probably be a fair-sized house, often of hewn logs. The old cabin would then serve as an outbuilding, and if it happened to have a dirt roof and lacked a foundation, it would quietly decay and finally be dismantled to make room for a modern building unless spared for sentimental reasons.

Thus, many of the old originals have disappeared. A beautifully preserved pioneer home of perfectly preserved logs was torn down because the owner, a farmer, needed the site for a larger building. Another fine old log house

Native village, Lillooet. 1960

By the Old Cariboo Road, north of Lillooet.

Abandoned house near Bridge River village. Once a home surrounded with apple and apricot trees, the old building now settles into the ground. 1973.

was dismantled presumably because it looked odd beside the new stucco house with picture windows. Yet another picturesque building of character was destroyed for no apparent reason — I think it interfered with the view. Other log buildings, chiefly barns, have had their roofs collapse under heavy snow — an understandable catastrophe, for shovelling snow off a barn roof is a formidable task. And fire, of course, is a notable destroyer of wooden buildings, having removed many fine landmarks through the years. In 1971 a forest fire swept in off the mountain and completely destroyed a native village at Lillooet.

So every year sees a reduction of the log buildings of British Columbia. In this age of rapid transportation the little log country schools are closed and soon become vulnerable to the weather. In any case log schools are out of fashion, as well as woodpiles; and school boards prefer to haul slick trailer schools back into the bush where necessary, and fuel them with propane gas. The modern trend has also affected some house owners to the extent that they have covered their lovely logs with patent siding, thus degrading the character of the landscape. This monstrous act of camouflage was explained to me by a log

Big Bar Creek, west of Clinton. 1960.

house dweller of Scandinavian descent — he confessed that in his case something had gone wrong in the building, some of the cornering was botched and he couldn't bear to look at it, so he covered the whole house with siding. Fortunately most of the log work of Scandinavian settlers is of such high quality that it stands proudly unadorned, causing no blushes to the builders.

Another trend at present evident in the country is the reroofing of old buildings with metal roofing, particularly aluminum. The harsh highlights of these flashy roofs are distracting features in the rural landscape, but let us be thankful that they have preserved many log buildings which would otherwise have succumbed to the weather. Metal roofing sheds snow, thus relieving the farmer and rancher of the onerous task of shovelling off a barn roof — or building a new one if he neglected to do so. The traditional roofing materials, shakes and shingles, were

George Elmes' house at Anglemont. Built around 1930, the house has a shingle roof, and upper floors are supported by hewn beams. Logs squared with broadaxe on the inside. 1973.

effective when properly maintained and had the advantage of harmonising with the landscape. A log building so roofed was an authentic component of the interior scene through all the days of gold rush and early settlement. After the newness of the peeled logs had mellowed, and perhaps the roadside dust had powdered them a little, a log house took its place in the landscape almost as a natural feature, like the rocks and the trees. The interior valleys and plateaux were sprinkled with such houses, many of them situated in the most hostile locations, for some of the builders were certainly over-optimistic regarding the potential of the soils they were locating. Such homesteads were eventually abandoned, the little groups of log buildings and corrals and straggling fences left to weather and decay, and become lichen-covered like the rocks and the trees. Lack of irrigation water, gravelly subsoil and unproductive locations defeated the pioneers; the larger ranches absorbed their homesteads but some of the old houses, with their unquenchable lilac bushes, are yet landmarks on their cattle ranges.

It is a strange fact that log and rail fences sometimes have a longer life than the solid log buildings built by the same hands, for once a fence is established in the cattle

Snake fence of jackpine logs. Chilcotin Plateau, Chezacut. 1968.

Chase Creek, near Falkland. 1962.

Abandoned homestead south of Kamloops. 1964.

Snake rail fence reinforced by aspens at Kersley, south of Quesnel. 1954.

country it has a good chance of being maintained, either in its original form or in a fresh pattern using the same materials. Trees grow up along fence lines, and their presence in a snake fence acts as permanent reinforcement to the rails, making the barrier much more effective and valuable. Some very old log fences, thus strengthened, continue to serve the rancher although deeply furrowed with age. They decay gracefully, these old log fences, weathering to a wrinkled old age; lichen-covered like the rocks, they straggle over the sagebrush-covered flats of the drybelt, and encircle many deserted homes.

The photographs in this book were made over the last twenty-five years by one who would much rather live in a log cabin with a dirt floor than exist in an apartment with wall-to-wall carpet and air conditioning, an unnecessary declaration, I suspect, to the reader who might think the work overburdened with ruins. Yet these ruins are our last tenuous link with the pioneers, and, unlike the citizens of the Old World, we have not our thousand years to ponder them. Stone is so much more durable than our wooden relics.

Relic by Monashee Road near Cherryville. 1962.

Dirt-roofed cabin with Russell Fence, Hat Creek. 1962.

Black bear and his tracks.

THE PRIMITIVE TOUCH

Temporary shelter was obviously the reason for a lot of the small log cabins situated in the more thinly settled parts of the province. Trappers, prospectors and ranchers built many intended for occasional occupation, and naturally these would be of the simplest construction and very small. Without exception they would have dirt floors and, less frequently, dirt roofs too — it depended on the materials available. The general rule was dirt roofs in the drybelt and cedar shakes in the moister forests where the conditions fostered the growth of that tree. Some trappers built a series of cabins on their trap lines; ranchers built them on their distant wild hay meadows to shelter a haying crew for a few weeks in summer, and a couple of cowpunchers feeding cattle in the winter months.

Prospectors in their private summer retreats in the mountains built cabins for seasonal occupation and for the storage of their equipment when they were reluctantly forced back to civilization for the winter. As these lonely cabins were commonly tenanted by mice and packrats, the prospector would hang his perishables — like flour, if he had any left — with a long wire to the ridge log to defeat those animals. He would probably attach a note to the door, before he left, addressed to the world at large, offering sanctuary, with a request that the place be left tidy and some dry wood be split to replace any used.

These primitive cabins served their purpose very well, and made snug retreats for those accustomed to living without benefit of furniture. Some boxes nailed to the walls made serviceable cupboards, hay on the floor a comfortable bed if the tenant disdained the luxury of a pole bunk, and a camp stove with a few lengths of stovepipe introduced real comfort for winter occupation. The smell of the interior of a log cabin thus warmed was the smell of home to the cold winter traveller who, once within its walls, was unaffected by the weather.

It was a simple job to build a log cabin where there was suitable timber available. A single man equipped with axe, saw and shovel could provide himself with a comfortable winter home in a very short time. I knew a trapper once, in the Chilcotin country, who wintered in such a cabin which lacked even the convenience of a camp

Pioneer's cabin at Deep Creek near Enderby, 1956.

Tiny window in outbuilding, south of Kamloops. 1964.

stove — he was in a remote area and had backpacked his supplies so he couldn't take in luxuries. He overcame the deficiency in a very neat manner by contriving a fireplace of rock and a wooden chimney in one corner of the tiny building. Slabs of rock protected the logs from the fire, and the chimney was a three-cornered affair, built with small, short poles laid log fashion and lined with clay. The logs of the building were small, chosen deliberately for easy handling, and the bark had not been removed as is generally the custom in more permanent constructions. His cornering was the common saddle-notch, the standby for rapidly — and roughly — erected buildings. It is made with an axe only, by chipping out a "saddle" in the under log and a notch in the upper log to fit it. This trapper had dug out his floor a foot or so, thus lessening the log work and rendering his den more cosy. The chinks between the logs were well stuffed with moss, and the roof was sturdily made with closefitting poles covered with an adequate layer of soil. He had built a bunk of poles which served him also as a table, and out of some boxes contrived a door, or rather hatchway, for it was very small. The hinges were strips of rawhide. There was no window.

I spent a night in this cabin in early winter, sleeping on the floor, and found it comfortable. The trapper made skilful use of his fireplace, turning out nicely-browned flapjacks. I subsequently learned that he wintered without hardship, but his enterprise proved unprofitable on account of the scarcity of fur in the locality.

Other Chilcotin log cabins in which I have lived for extended periods were equipped with both stove and window, and a real door sometimes as much as five feet high; but the dirt roof was the rule, as was the dirt floor.

Outbuilding and cabin, Dog Creek.

Early settler's home near Falkland.

Cabin in the woods at Duncan Lake. 1965.

Top to bottom:
Cabin at Two Springs, Hat Creek Valley. Near Alexis Creek. Deep Creek, near Enderby. Old Fort Steele.

Cowpunchers and ranch hands accepted the dirt floor without question, but miners, apparently, like things a little more civilized, as I found out later when building log cabins with a party of miners and prospectors "back in the hills", as they termed the Coast Range. These fellows had brought in a whip-saw, and after constructing a saw pit, went to work sawing out planks with which they floored the cabins and constructed tables, benches and doors. It seemed an awful lot of work for a mere summer's residence.

That was many years ago, before the helicopter had made access to remote places so easy, and rendered the prospector's friend, the pack-horse, obsolete. Modern communications have disrupted the leisurely lives of the old-timers, and introduced problems which they never had to wrestle with. The tenant of a log cabin never had to worry about the plumbing going wrong, or the wiring, or the furnace quitting, or birds flying through the windows — the windows were too small to attract birds — or the paint peeling, or any of the hundred other items the modern householder has on his mind. If the dirt roof leaked he avoided the drops and they were absorbed in the dirt floor. But a well made dirt roof didn't leak — unless you left an unusual amount of snow on the roof and were caught in a sudden thaw. I lived in a dirt-roofed Chilcotin bunk house for a year and experienced no leaking, nor any other problems either, except the occasional nocturnal disturbance caused by a packrat. Fire, of course, was a possible hazard, but if it happened the loss would be trifling, and the building could quickly be replaced. The assessor never came round to see what you'd been up to in the way of additions and improvements of a taxable nature. Altogether the life of a log cabin dweller, with all these advantages, had much to recommend it. The very look of these small buildings rule out all notions of hurry and tension applied to the inhabitants who once dwelt within their friendly walls.

Canoe Creek, west of Clinton. 1971.

Tatlayoko Lake. 1965.

Above and below, log houses at Chezacut on the Chilcotin Plateau.

Right, Tatlayoko Lake, in the Far Chilcotin.

RELICS OF PIONEER DAYS

The missionary, Father Pandosy, established his Okanagan mission in 1860. With materials obtained from the neighbouring bush he erected log buildings which can be seen today a short distance from Kelowna, where they represent perhaps the oldest examples of log buildings existing in the interior. The construction of the little mission is very sturdy, and is obviously the work of an experienced builder who intended his work to last. The logs used are fairly large and hewn flat with the broadaxe, after the fashion of the early log men who tried to achieve as smooth an exterior as possible on their dwellings. Dovetailing of the corners proclaims the competent workman, and the evenly dressed logs, pride of workmanship. It is no wonder that these walls are standing today. The buildings themselves were recognized as historic relics in time to protect them from the processes of natural decay which work chiefly from the roof and the ground, so re-roofing and attention to foundations ensured their continued existence.

The mission became the nucleus of a settlement that saw the start of agriculture and fruit growing for which the valley later became famous, and the well made buildings doubtless set the pattern for some of the settlers' homes. That the standard was not always maintained is apparent in later buildings that have survived. Round Prairie School in North Okanagan is a relic of the first little agricultural community that subsequently expanded into the municipality of Spallumcheen, but its logs show none of the careful workmanship that distinguishes Father Pandosy's Mission. It was built in the summer of 1885 as a community project by the local settlers, some of whom were veterans of the Overlanders Expedition of 1862. The school stands in an open pine grove on the McKechnie farm a couple of miles north of Armstrong, its stout logs resting dry and secure under a renewed roof of cedar shakes. It is an example of the log buildings that were erected when sawn lumber became available in the interior, and dirt floors went out of favour, at least with the married settlers. It was a considerable advance on the cabin. Logs for the walls, and perhaps the joists, lumber for the floor, ceiling and inside finish, became the rule in house construction. The amount of lumber used was at the

Above, the dovetail cornering of Father Pandosy's Okanagan Mission, built in 1860.
Opposite, other buildings at the Pandosy Mission near Kelowna.

Right, home-forged calipers for marking logs, a prospector's mortar and pestle, and old bottle.

Below, settler's original cabin, with dirt roof, in the Salmon Valley, near Falkland. 1960.

discretion of the builder who would naturally not buy it if the material could be obtained from the bush for nothing, except for labour, which in those early days represented very little money. The broadaxe was then in common use among the settlers who manufactured accurately hewn joists as well as barn and bridge timbers, and later, railway ties.

But Round Prairie School exhibits no particular refinements of construction. It would probably have been regarded as a rough job by the axemen of the day, and the settlers who built it were not unduly concerned with demonstrating their skill. Its chief interest lies in its undisturbed situation in a segment of the original landscape, a small log building in a pine grove on a sandy hill, doubtless little changed since the first settlers' children of the sparsely occupied valley made their reluctant way towards it.

Round Prairie School, built in 1885

Samples of corners from several old cabins, and the froe, or frow, used for splitting shakes. The tool has a knife edge for splitting and a broad back for striking with a wooden mallet.

More corner samples, and the broadaxe, the tool that made it all possible.

A relic of very different character is Cottonwood House, in Cariboo, on the Quesnel-Barkerville road, which is said to date from 1864 when it was established as a roadhouse on that much-travelled route. It is a substantial, two-storey building of well fitted hewn logs with corners neatly and expertly dovetailed, a handsome prototype of the buildings that later became familiar throughout the Cariboo, especially along the main route of travel, the Cariboo Road. Two-storey log houses were a great improvement on the lowly cabins of the settlers and miners, and wherever they were built they gave an impression of permanency to the ranches and settlements which they enhanced. They were built where suitable timber was handy, and sometimes in spots where it wasn't. Some old houses in the drybelt sit in the open, on the floor of treeless valleys, indicating that the logs were hauled by teams, perhaps over many miles, from the nearest timbered hills.

A Cariboo relic — Cottonwood House, on the road to Barkerville. This roadhouse is reputed to have been built in 1864.

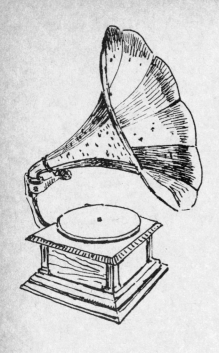

Much of the log work of the pioneers was remarkably accurate, carrying the stamp of the proficient craftsman. Anyone handy with the axe can chop out the simple saddle notch and lay logs successfully, but the dovetail corner demanded a certain amount of skill and experience. Dovetailing was used on both round and hewn logs. In the case of the former the two ends of the log would be squared — the log being set up on notched skids for the operation — using a spirit level to mark the ends to ensure a perfect vertical face on each. If the log was to be hewn, a well chalked line was held tightly against it, then drawn back like a bow-string and snapped, which transferred the chalk on to the timber. After deeply scoring the log with an axe, the broadaxeman hewed to the line, producing a more or less flat face on the log depending on his skill with the tool. Broadaxe work added a tremendous amount of labour to the construction of a log house without enhancing the strength of the building, but it was probably practiced so frequently because of the inate desire of the

Creighton Valley near Lumby.

craftsman to do a good job. Hewn logs looked more professional, removed the building a little from the rawness associated with the standing bush; the logs had been converted into timbers, and the house, hopefully, from a distance might even be mistaken for a frame building with dressed siding.

Many of the old houses that survive bear the marks of this heavy work with the broadaxe. They were certainly not thrown up in a hurry, being, like Father Pandosy's mission buildings, built to last; and they were rarely abandoned because of deficiencies of construction. The choosing of a poor location seems to be a major reason for the abandonment of pioneer homes. British Columbia's climate and conditions are so variable that it was impossible for the settlers to correctly evaluate a prospective homesite on a brief acquaintance with the new country. They built in the drybelt in hopeless situations and were dried out; they settled in the green bottoms of the North River and were flooded out; they pioneered in side valleys at altitudes that precluded most possibilities in agriculture, and they attempted lakeside clearings on ground that was too steep and rocky to provide any but the most precarious of livings. Consequently their sad relics are frequently to be found in remote situations, off the beaten track, in overgrown clearings and deserts of sagebrush and prickly pear.

Top, farm house with boarded corners, near Salmon Arm.

Centre, school at Tatlayoko Lake.

Pioneer cabin, Semlin Ranch, Cache Creek. 1955.

Opposite, top, abandoned pioneer home in the North Thompson Valley near McLure. Bottom, old log house, D.C. Jones ranch on Texas Creek Road south of Lillooet.

But the more successful, or luckier, settlers developed their holdings into fertile farms and fine ranches. Their locations became permanent and the log buildings were maintained, especially the barns and outbuildings which never outlived their usefulness. Under new roofs they remained sound and solid, enduring and picturesque examples of the settlers' industry and skill. The log houses though, were difficult to remodel and adapt to modern conveniences. Families outgrew them and built new homes with basements, which the log houses generally lacked. Not many of the pioneer homes remain today. Some have been torn down recently, and over the last ten or fifteen years many have disappeared, but a few are still lived in and, hopefully, will be occupied for some time to come.

Above left, Hudson's Bay Company hewn log buildings, Fort St. James.

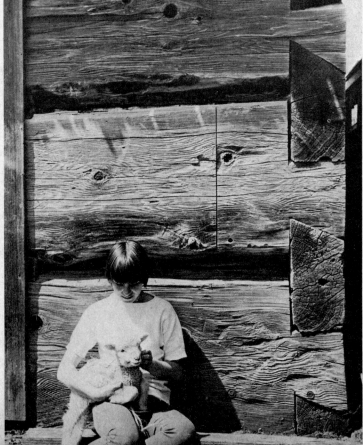

THE NATIVE VILLAGES

Probably because they are mostly situated in rural areas, the native villages of the interior — rancheries, they call them in the Cariboo — run heavily to log buildings, or rather, did, for over the last decade the new housing program for Indian reserves has replaced many of the old houses with new frame dwellings. Before this happened the common impression of a native village was of a cluster of log houses gathered near a church which would nearly always be of frame construction, and very likely painted white. Presenting a strong contrast to the dwellings, the church, sitting on the most prominent site available, advertised the village from afar. These interesting landmarks were all of different design, easily recognizable by a wanderer of the interior who could name the village if shown a picture of the bell tower or spire. Not so with the houses built with the traditional logs. Who could say whether this or that house was in Lillooet or Anahim; Fountain or Sugar Cane or any other of the interior villages?

Hewn or in the round, these village log houses were almost without exception expertly built. Proficiency with the axe and broadaxe was universal at the time of their construction, which must have been around the turn of the century. Some of the houses were older than that of course, much older, judging by the weathered logs and general condition of the building. A woman of Alkali Lake village who had been doing some remodelling in her log house told me that she found a newspaper of 1907 in the walls. Her house, like most of the others at Alkali Lake, was in fine condition with no sign of deterioration of the log work, and was obviously of a later period than many of the old Cariboo relics. Now the gradual change to frame buildings is altering the character of the village. Progress, it seems, is inevitable, and log houses are considered old-fashioned and out of date, but the arguments in favour of them are not entirely sentimental. "I like my log house," the woman of Alkali Lake told me. "I will be the last to move into a frame house. The people in the new houses are cold in the winter, but I keep warm in mine with just the cookstove going."

If the new houses improve the living standards of the villagers they will do little for the appearance of the

Above, log barns, rail corrals, Merritt country.

Below, log church at Lillooet.

Left, Fountain Village, near Lillooet. 1953.

Below, log church at Niklepam, west side of Fraser Canyon between Lytton and Lillooet. Small cabin alongside was overnight accommodation for preacher when village was only accessible by trail.

settlements, for no wooden building wears better or fits more harmoniously into a landscape than a log house. The villages will no longer be the same if the modern trend continues. Already many replacements have been made, and villages which a few years ago were predominantly of log construction are now greatly modified in appearance.

Of all the native villages none has a more spectacular situation than Fountain in the Lillooet country. Nestled in a high hanging valley in the Fraser Canyon it is overtopped and dwarfed by precipitous mountains with sharp peaks. It is far above the old Cariboo Road that was built in 1861 from Lillooet to Kelly Lake, and still lies like a crooked thread across the ridges and gullies of the great gorge. Fountain village is perched half way between the river and the mountain peaks, a curiously remote fastness for a people that have always fished the river for its salmon. It is an old village which used to contain log buildings and relics of buildings seemingly as old as any in the Lillooet country. The outlying little farms and homes, too, that occupied tiny benches and hollows in the canyon below the village produced examples of early log work in their houses and outbuildings.

Bridge River Village near Lillooet.

Log church near Lillooet. 1962.

Fountain is an eyrie far exceeding in altitude all the other canyon settlements. Its near neighbour, the old deserted village of Bridge River, is near the bottom of the gorge, only a few hundred feet above the Fraser at its confluence with the Bridge, and has been abandoned for many

years. An old man who was born in the village in 1897 told me there was no water at the site and all requirements had to be carried laboriously up the steep gravel hill from the river. Apart from this handicap the village was beautifully situated and has been much admired by travellers

House with spiked corners near Fountain Village.

Above right, the village street, Lillooet native village. Completely destroyed by fire in 1971. 1960 photo.

Above, left, deserted native village, Bridge River, near Lillooet. 1969.

Old hewn logs in Bridge River Village. 1967.

looking down on it from the canyon road, a thousand feet above it. There is not much left of it today. Only a few of the log houses remain, and the church burned down a few years ago. A burial ground with many graves, overgrown with sagebrush, looks down on the river from the edge of the bench.

The native village on the outskirts of Lillooet was completely destroyed in an hour in 1971 by a bush fire driven by a high wind. It too was an old village, one of the interesting sights of Lillooet, with its sturdy hewn log houses and handsome church; an ornament to the local landscape which can never be replaced. Many log buildings, however, do remain in the Lillooet country, particularly on the west side of the canyon below the old town where, for forty miles, every little bench with a creek available shows signs of present or past occupation. The extreme ruggedness of the terrain precludes the normal village arrangement; the homes are scattered and hidden in the folds of the canyon, but the local church indicates a location roughly central to which the people toil, either steeply uphill or down. Nyshakup, Niklepam and Barcheesan, which is across the canyon, are tiny churches that seem crowded with a dozen worshipers, but they exude the very spirit of the mountains. Niklepam church is of ancient hewn logs with dovetailed corners but the bell tower is frame and shingle-covered, a combination which in no way impairs the canyon landscape.

Alkali Lake Village.

42

In the Chilcotin country log houses were commonplace and the natives used them extensively for their village homes. The large village of Anahim used to present a fine appearance with its many houses and church clustered on a valley bench which is backed by a crescent of hills capped with vertical rimrock, but its smaller neighbour, Stone — or Stony, as the locals call it — is hidden away from the world, across the Chilcotin River, and can be reached by those who know the way by an inconspicuous side road. Farther back in the Chilcotin are the villages of Redstone — which sits near a cliff of red volcanic rock — and Anahim Lake, where the art of log house building is still practiced. With so much good material available it is probable that the log house will survive for a long time to come at Anahim Lake. A good stand of close-growing jackpines furnishes perfect building logs, even in size and tapering very little, with the added advantage of having no limbs. When matched and built into a house they present a very smooth appearance. The village houses and the children's dormitories operated by the Missionary Sisters are good examples of this smooth log work. Dovetailing of the corners has not been attempted — it is rarely practiced nowadays — but a tighter join than the saddle-notch has been achieved by an interlapping square join.

There is little doubt that on the Chilcotin plateau log building will continue to flourish and villages like Anahim Lake will retain to some extent their frontier

Anahim Lake Village. 1964.

Spahomin Village, Nicola country, near Douglas Lake. 1968.

Shuswap native village, Upper Columbia Valley. 1960.

House of hewn logs, Sugar Cane Village, near Williams Lake. 1963.

Alkali Lake Village. 1965.

Sugar Cane Village. 1963.

flavour, which other settlements are gradually losing. Some of them sit in the open far from suitable timber. Their old log houses were built in more leisurely times when the sawmills were not competing for every sound stick of timber over a six inch top. Hauling logs was then a winter occupation with teams and sleighs, a downhill haul from the nearest timbered hills which in some cases were miles away. The quaint little village of Toosey near Riske Creek occupies a hollow in the treeless range, and must have hauled its logs for some distance. In the Douglas Lake country Spahomin, too, has old log buildings in an open landscape of grassy hills. The old church of Spahomin village, a frame structure, is said to date from 1884, and judging by the looks of the log buildings remaining they could be about the same age.

The Thompson Valley villages have been losing their log houses for years, although a fair number remain in scattered locations in the neighbourhood, chiefly farm and ranch homes which are now unoccupied. The river-bank villages of Squilax, Chase, and Shuswap have changed in recent years, but in each case the church retains its prominence as a familiar landmark. The same can be said of Deadman Creek village where the immaculate white church with its red roof used to pose such a startling contrast to the weathered logs of the houses. Bonaparte village near Cache Creek still has an old-time look despite the new houses. The highway divides the village from its large and interesting cemetery which contains the graves of persons born in the eighteenth century, already old people when they saw the Cariboo Wagon Road built through the village in the eighteen-sixties.

Bonaparte Village near Cache Creek. 1960.

FENCES AND CORRALS

When I was working at Chilco Ranch in the Chilcotin Valley in 1928 a story circulated in the bunk-house that a young man of nearby Big Creek had once cut 400 rails in an hour. That the story was accepted without question is remarkable testimony to the quality of a stand of Chilcotin "rails," where the close-growing trees make such a performance conceivable. No one doubted the achievement, but there was no indication that any of the hands desired to emulate the prowess of the Big Creeker. They were content to produce rails at a more leisurely rate, and while they undoubtedly did cut many rails in nine seconds or less, that rate would rarely be sustained for more than brief periods. Ever since I heard the story I've had a mental picture of the young man, in the pride of his strength, laying about him lustily with a sharp axe, two chops for the butt, one for the top, stepping swiftly to the next tree and laying it down in seconds. I see the trees falling like ripe grain before the sickle, the rails littering the ground, and the splendid young man stepping over the tops, wading into the standing forest with undiminished vigour — nine seconds to the rail, 400 to the hour. The poet Grey would have approved: "How bow'd the woods beneath their sturdy stroke!"

Above, a well-worn, but still functional log fence.

Opposite, rail fences at Ryder Lake, lower Fraser Valley. 1969.

There would be no limbs, of course, to hinder the axeman. What rudimentary branches that started with the sapling had shrivelled to mere whiskers in the dense stand of growing trees. They would be rubbed off later with handling the rails.

The proliferation of rail fences in the interior was the natural outcome of such stands of raw material, both jackpine and fir, that were to be found in the areas taken up by prospective ranchers and farmers. The material was available to the settlers without cost; it could be put into effective use in an unmanufactured state, and was the perfect answer to many essential needs. The fence pattern of the interior was established by the early settlers with designs so inherently sound that all of them are to be seen in use today. In many cases the material contained in rail fences at the present time is the product of the pioneer axeman who spent many of his winter days in the bush cutting and splitting rails. Not always were stands of rail-size poles convenient, so large trees had to be split, and it is these

Dodgeway on right permits person to walk through fence, but stops livestock.

Post and Rail Fence, Turtle Valley.

split rail fences that became a feature of the farming areas where more fertile soil conditions produced heavier stands of timber. The farmer, too, cleared land, which meant making a clean sweep of the timber on it and utilizing it all, so much heavy stuff would find its way into the fences in the shape of split rails. And split rails are very durable. They account for many of the old-time fences seen in the farming districts today. The posts have rotted and been replaced perhaps a dozen times, but the old rails endure and go back into the fence every time it is rebuilt.

The posts were the weakness of the rail fence. They rotted off below the ground level quite rapidly, making upkeep of a farmer's fence system an annual chore. Some farmers charred the butt ends of their posts in the belief that the treatment would prolong their usefulness, or they seasoned them for a year before they planted them. Some believed a split post would last longer than a round one, but all agreed a cedar post would outlast any other, so cedar posts were favoured, and became an important source of income for those farmers with stands of cedar. There is a wood which will outlast cedar in the ground but it is scarce and not found in any quantity in the farming districts. The wood is juniper, indigenous to the drybelt where it spreads its crooked limbs over the rocky hillsides of the Thompson Valley. Jim Allan, rancher of Deadman Valley near Savona, told me he had juniper posts standing after sixty years in the ground. A whole fence had lost only two posts in that time, broken off by a couple of bulls

Above, Russell fence on Carson Ranch, Pavilion Mountain. 1952.

Below, rancher's snake rail fence near Alexis Creek. 1960.

fighting. These posts supported wire — for rails are scarce in Deadman Valley — and most of them were so small as to represent mere stakes. It is not easy to find juniper that will make a sizeable post.

Why did the farmer bother digging post holes when other types of rail fence were just as efficient? The other types occupied more ground which on good agricultural soil was an important consideration; they interfered with cultivation and harboured weeds. The rambling snake fence was good enough for the pasture where the cattle and horses grazed out all the corners, but for dividing his cropped fields the farmer likes a straight line to which he can plow close. The post and rail is essentially a farmer's fence. The materials can all be had free with the exception of a few hanks of pliable wire to attach stakes to the posts to form a slot into which the rails are inserted — for that is the traditional method, stronger than the more recent expedient of spiking the rails to the posts. Its general utility established it as a favourite in widespread areas. Well made, it presents an inpenetrable barrier to the discontented cow forever looking for a hole through which to thrust its head, and it can be climbed over and also sat upon and leaned against by tired hired men.

This log fence at Chase Creek was built in the 1920's. 1969 photo.

The rancher, however, has little interest in the post and rail. His use of it is nearly always confined to the construction of corrals where heavier material is not so easy to obtain. The rancher will often go to a lot of trouble digging post holes and setting enormous posts for the gates and chutes of a complicated system of corrals, but for his large pastures, his drift fences, his protective fences around range hazards such as dangerous bogs, and the fences around his wild hay meadows, he prefers to avoid digging post holes — unless he can find juniper posts that will last for sixty years.

This disinclination on the part of the rancher for the robust exercise of digging post holes is not due to a lack of enthusiasm for physical exertion. It is rather dictated by the nature of the ground on which he centres his activities. Ranching country is traditionally wide and open, of a nature that precludes farming or any intensive cultivation on account of factors such as dryness of climate and shallow and rocky soils. The Cariboo and Chilcotin plateaux are examples of such unfavourable terrain. Much of the land is boulder-strewn with as much rock hiding underground as exposed above, so it is quite impracticable for setting fence posts. Fortunately, the amount of

An old fence repaired with a new top log, in the Bridge River Valley, near Yalakom River. 1967.

rail material growing there is inexhaustible, and is utilized in fences of various patterns, none of which depends on planted posts. There are two main types: the snake fence which rambles over the terrain zig-zag fashion, sustained on the principle of mutual support; and the A type which uses crossed stakes to carry the structure. Each particular pattern has a definite reason for its existence, generally more substantial than the mere whim of the rancher. Availability of material was, of course, the main consideration. If it was plentiful and handy a snake fence of some kind was likely to be built, if scarcer and hard to come by an economical Russell fence would be suggested. This Russell fence straggles across much of the Cariboo country. It is the hardy perenniel that gives to the country a special character. Many times, indeed, it is a sorry sight to the consciencious rancher who views its sagging rails and splayed stakes with concern knowing that a determined or disgruntled "critter" would find it but a feeble obstacle. Luckily the Cariboo range cattle respect rails while the hold wire fences in contempt. The Russell fence frequently fails to maintain its stability because it was built with rails so light that they soon sag, and the binders sometimes slacken, causing the stakes to spread. An added hazard in the days of horse transportation was the habit of the passing teamster of going to the fence for wire for some minor repair or to wrap a loose wagon tire. This caused the collapse of a panel and hastened the deterioration of the fence.

Sturdy Russell fences, however, were and still are built and maintained by numerous ranchers who find it very economical in both material and labour, for, despite its complicated appearance, it is swiftly and easily erected. Old photographs of the Cariboo show the fence in use in the early days; it is certain to continue to be a familiar feature of the ranching scene.

The Russell fence requires five rails to the panel, the snake rail fence eight or nine, or perhaps more, depending on its height. The Russell fence is built on the straight, but the snake fence zig-zags, so the rails cover less distance than those of the former. These advantages ensure its continued popularity. The snake rail fence is rarely built nowadays, but where old ones remain in reasonable shape it is generally worth the rancher's time to maintain them. The large quantity of material required for its construction rules out a modern revival of this type of fence.

It is different with the log fence though. Usually built in the same zig-zag fashion, it remains a prominent and picturesque feature in ranchland scenery. Judging by the ancient relics of log fences to be seen throughout the interior it was used in the earliest times. It is still being built today, the sturdiest and most enduring wooden fence ever

Above, a flimsy Russell fence in the Cariboo, near Jesmond. 1965.

Opposite, a farmer's lane with post and rail fences near Enderby. 1959.

Below, repairs are needed on this fence at Tatlayoko lake. 1965.

Less common way of connecting panels of log fence with short blocks. Turtle Valley. 1968.

Stake-and-Rider fence made with stout poles at Bear Creek, near Chase. 1964.

devised, and also one of the easiest to maintain. The occasional replacement of a top log which is most exposed to the weather, is all that is necessary to keep it in serviceable condition. It required a lot of labour to build, especially in the early days when the logs were felled by axe and crosscut saw. Many of the old fences were made with large logs, three of which to a panel raised a sufficiently high barrier. The bottom log would be set well clear of the ground on a rock or wooden block, and thus elevated would defy the weather for half a century or more. These large-log fences were nearly always the work of old-time cattlemen who didn't have to consider the value of the massive logs they put into them. The modern rancher is content with smaller stuff, but I knew one — albeit of forty years ago — that liked large logs, especially for the top — "to hold it down," he said. It was a bit of a struggle getting them up there, but they served the purpose very well.

This rancher had a log fence eight miles long which was built in the nineteen-twenties when the contract price for the fence was 400 dollars a mile. This fence enclosed some range and wild meadows on the jackpine plateau of the Chilcotin country, and was about twenty-five miles from the home ranch. It was five logs high, chiefly jack-

pine. The rancher used to wonder how so many cattle of a different brand got into his pasture although he could find no breaks in the fence. I could have enlightened him but didn't. I had previously worked for the other brand, wintering cattle on their plateau meadows with an old cowpuncher. Tex had charge, and when spring came and he considered the grass on the range was good enough, we drove the bunch out. A number of them kept returning to the meadow where we were feeding a few cows. Tex was exasperated and vowed he'd "fix those critters." This time we drove them in a different direction several miles through the jackpines until we came to a log fence. "So and so's pasture," Tex grinned, as he dismounted. We opened a panel in the fence, drove the bunch through, and closed it up. We saw no more of them.

That is a unique and convenient feature of the snake fence — it can be opened and closed with little trouble at any point, unless, of course, built of large and heavy logs.

Other structures built of logs and rails are the corrals of the ranchers, scattered through the cattle country adjacent to the ranch buildings, on the open range, or in sheltered hollows in the bare hills. Every ranch needs these installations to facilitate the operations of branding, vaccination, dehorning, sorting and, nowadays — since the cattle truck rendered the beef drive obsolete — loading. The simple round corral is commonly built of logs, the larger the better, peeled to inhibit early decay, and raised much higher than the ordinary fence, five or six

Farmer's post and rail fence near Lumby showing original method of attaching rails with stake wired to post.

This old Russell fence has seen better days. In the Chilcotin Valley, near Redstone. 1966.

Rail fence and aspen trees — a typical Cariboo scene, near 100 Mile House. 1954.

Above, left. Four logs high make a stout corral for rancher Bill Philip in Deadman Valley.

Above, right. Corrals at Spahomin Village, Douglas Lake country. 1969.

Gate and fence of peeled poles in the Chilcotin jackpine forest, Chezacut. 1965.

Corral and log barn on the old Meason's Ranch, near Dog Creek. 1965.

Rancher's corral in Deadman Valley. 1956.

logs high or perhaps more, to form an insurmountable barrier to animals not accustomed to confinement. More complicated corrals require the use of both logs and rails in a system of pens, enclosures and chutes and the intervening gates which result from the special needs of the rancher or perhaps his talent for innovation. Gates may be stoutly made with peeled poles and hung to tall posts sunk deeply in the ground. Posts and heavy rails are used for the narrow chutes through which the cattle are passed singly, in fact everything about the corral except the ironwork of the dehorning chute can be got from the standing bush.

These log and rail corrals are interesting diversions, especially in districts where wire fences predominate. Like log buildings, they rather enhance the landscape; they harmonize with the dusty, short-cropped range and the solitary trees. Bleached wood looks well in the drybelt, while the invading wire can but impair the scenery. Fortunately, the rancher is not likely to run out of material for his fascinating productions, and with that much maligned machine, the chain saw, at his disposal, his works in the shape of fences and corrals will continue to prosper.

The rancher's favourite fence. Rimrock never needs repairs. 1952.

THE SCANDINAVIAN WAY

The owner said proudly, "built in 1897 and been in the family ever since." An exceptionally solid and attractive job, I thought; large logs hewn almost as smooth as dressed lumber, the corners perfectly dovetailed and pegged, the owner said, with wooden dowels, the whole structure, in fact, appearing as square and sound as when the logs were first laid. It is a two-storey house, ten logs high to the eaves, and the same carefully hewn logs carried up to the peak of the gables. Chinking of such close-fitting logs can hardly have been necessary and must have been carried out solely for the sake of ornament.

This fine house stands in the little settlement of Hagensborg in the Bella Coola valley about nine miles inland from the head of the coast inlet, Bentinck Arm, and is an example of the particularly fine log work to be seen wherever the original settlers were of Scandinavian descent. Hagensborg was settled by Norwegians but Finnish communities also contain samples of similar workmanship showing the same skill and attention to detail, and the obvious pride in the job that has made many of these old buildings outstanding landmarks long after their builders have passed away.

One such landmark is a remarkable log house on a hilltop site at Notch Hill, the little settlement that got its name from the steep railway grade which still slows traffic on the Salmon Arm-Kamloops section. It is a nine-room, two-storey house, 27 x 34 inside measurement, 20 logs high to the eaves with a central log partition, and interlapping square corners which have been covered with decorative corner boards. The logs are hewn both inside and out, and the broadaxe work is quite exceptional. No smoother job of hewing could be imagined as the score marks of the axe are entirely absent throughout the whole structure; the logs, in fact, are as smooth as boards and so evenly matched that from a short distance the log construction is not readily apparent.

The builders, August Nelson and Nels Sjodin, were Swedish timbermen, highly skilled and talented. They commenced building this masterpiece in 1909 and finished it a couple of years later. The logs were laid after the fashion in the Old Country, the process being recalled by an

Above, log barn at White Lake with space between two buildings roofed to provide cattle shelter. 1957.

Opposite, window in hewn log house at Notch Hill. 1974.

Norwegian style hewn log house, Hagensborg, Bella Coola Valley, built in late nineteenth century. 1964.

old-timer of Notch Hill who knew the builders. After hewing, the under log was tongued and the upper grooved to fit it and the joints packed with oakum, which made such a tight join that outside chinking, the usual way of packing a log house, was never required. All logs were pegged at the corners with wooden dowels. Besides being beautifully executed, this house has much architectural merit. It is a very well proportioned building, presenting a handsome exterior which owes not a little to the arrangement of the windows and their white-painted frames. Its excellent condition is a tribute to the builders and the present owners who have been attentive to maintenance. A preservative stain has recently been applied to the previously untreated logs, and all trim and woodwork painted.

Conventional interior finish has covered up all but a few square feet of the wonderful work of the broadaxemen. One wonders what kind of men these were, to hew so splendidly, striving for and achieving perfection, and then to hide the beauty of the dressed logs under common wallboard. Even the central hewn partition, if left uncovered, could have warmed and enhanced this interior, distinguished it with the simple honesty of logs; but perhaps that would have been too rustic for the builders, too reminiscent of the humble log cabin which to them must have signified the ultimate attainment of the unskilled.

Expert dovetailing in log barn at White Lake, Salmon Arm district. 1957.

Opposite. Even the sauna is of hewn, dovetailed logs in the White Lake settlement.

The British Columbia interior has several settlements where the work of the Scandinavian timberman can be recognized. Finnish settlements at Solsqua in Eagle Valley, and White Lake near Salmon Arm, contain many fine buildings and whole farmsteads where all buildings from a large barn down to the tiny sauna are of hewn and dovetailed logs. The work of the Scandinavian settler is characterized by the skilful broadaxe treatment, nearly all of their houses and many of their barns being constructed with hand hewn logs. In some of the houses, especially, the work of squaring the logs has been so expertly performed that the building presents a smooth exterior in which the individual logs have almost lost their identity.

I was admiring a house of this description at White Lake and was discussing its merits with the young housewife. It was a two-storey house, hewn and dovetailed, and I decided it was a first class job. The young woman didn't agree.

"Grandpa built it and now he can't bear to look at it," she told me. "There's a terrible mistake in it. Grandpa feels awful every time he comes here. He wants us to cover the whole thing with siding."

I scrutinized the house minutely but could find no fault in the workmanship.

"It's around the back on the upper storey," she explained.

We went around and scanned the dovetailing from top to bottom. The young woman looked puzzled. "I know it's here but I can't spot it today," she confessed, "but grandpa groans every time he looks at it".

A lot of the White Lake buildings were erected in fairly recent times, although before the days of the pickup baler which has influenced the design of modern farm buildings. The hay loft was of prime importance when the White Lake settlers built their barns, so the log walls were carried up high enough to provide ample storage space for hay above the stables. Settlers first went into the area about 1910 but most of the original homes have been replaced, and some of the later buildings were erected during the thirties and forties. Dovetailing was a feature of nearly all their buildings, indicating that the descendents of the pioneers possessed the old skills and could build as solidly as their forebears. They also retain the same modest estimation of their accomplishments. I once complimented one of these people on a building that had taken my fancy, but was mildly rebuked for my implied ignorance of such things. He looked sad and said it wasn't much, a rough job really, run up in a hurry. The logs in this case were left in the round, which is apparently sufficient reason for a Scandinavian timberman to belittle a job.

Opposite, top. Log barn at Malakwa, with the steep roof common in regions of heavy snowfall. 1968.

Opposite, bottom. Large barn of hewn logs at Solsqua.

Farmhouse of hewn logs at Solsqua in the Eagle Valley. 1964.

The Solsqua settlement at one time used to present a real "Old Country" appearance owing to the curious method of haymaking the settlers imported from their homeland. This was made necessary by the damp conditions of deep and narrow Eagle Valley where heavy dews and mists prevent or delay the curing of hay left on the ground. The Finnish method was to set stakes with cross-bars about the field and build up tall pillars of hay on the stakes, where, being clear of the ground, it cured rapidly. However, the pickup baler rendered this method obsolete, and Solsqua's Finnish flavour now depends on its log buildings — houses, barns and saunas. Like the White Lake specimens they are strongly and expertly built, but the barn roofs are steeper, designed to withstand snow which is unusually abundant in Eagle Valley. The amount of snow these Eagle Valley barns withstand is truly astonishing, and as the valley penetrates deeper

into the mountains the barn roofs seem to carry a steeper pitch to cope with the increasing snow load. But the ultimate answer to the terrific snows of the region was found by some Revelstoke district farmers who dispensed with walls and built A-shaped barns, all roof and steep as church steeples on which it was impossible for snow to accumulate.

Houses, though, were not purposely adapted to climatic variations. They were all fairly steep-roofed anyway, after the custom of the period, and when metal roofing was introduced it quickly became popular in the snow districts. Householders no longer had to shovel their roofs as the snow slid off the metal.

At White Lake not all the houses can be recognized now as log buildings, having lost their identity under exterior coverings of wood siding, stucco, and even patent

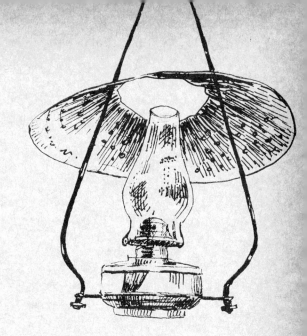

Expert workmanship in hewn log barn at White Lake. 1974.

imitation stone siding, indignities that are never inflicted on the barns. But some residents are still proud of their logs. Mr. and Mrs. Waino Nayki live in a lovely log house which was built by Waino's father, an immigrant from Finland, in 1921 to replace the original home which was destroyed in a fire. The young family lived in the sauna — fortunately a large one — until the new house was built. It is a fine, seven-roomed house of two stories built of hewn, well-matched cedar logs which were cut in a fire-killed bush and required no seasoning. All the joists are hand hewn timbers. The interior was originally finished with wallboard and wallpaper but Waino, while remodelling, stripped off some of this and exposed the beautifully hewn logs. Being a broadaxeman himself he dressed them a little finer and left them to grace two walls of the living room.

The Nayki house presents the usual smooth exterior with closefitting logs and perfectly dovetailed corners, albeit covered up with the decorative corner boards favoured by some Scandinavian builders — decorations which Waino thinks he might remove.

He told me that the close fit of the logs was achieved by hollowing out the under side of the upper log before setting it down on the rounded upper side of the under log. This involved accurately marking or "scribing" the upper log with a home-forged, calipers-like tool which left a level scratch mark along the log as a guide to the axeman who would hollow it out to the line, using the corner of a double-bitted axe for the job. It was usual for a man to work at each end of a log for dovetailing, each man staying with his own end, and a left-handed fellow being a decided advantage at one of them.

Hewn and dovetailed sauna. White Lake. 1974.

Opposite. The home of Mr. and Mrs. Waino Nayki.

Above, left. Barn with shake roof at Solsqua in Eagle Valley. 1964.

Above, right. Log outbuilding at Malakwa. 1968.

Below, left. Log barn, shake roof, at White Lake. 1974.

LIVING WITH LOGS

My friend, the late Gladys Adler of Six Mile Creek, could feel comfortable only in a log house. She was already an old lady when I first became acquainted with her and visited her Little Kingdom Ranch, hidden away, up a side valley on the west side of Okanagan Lake. Little Kingdom was, she said, her "perfect location", the sixth, I believe, and last of the ranches she had owned in the British Columbia interior. She was an old-timer and had lived in several districts, including Upper Hat Creek, Deadman Valley, Walhachin, and the valley of the North River, as she called that branch of the Thompson; and the term "ranch" which she used in describing these properties signified large areas of rough country surrounding a few log buildings and a corral.

The Six Mile Creek ranch followed the pattern but the log house was just about uninhabitable. It was very small, and the bottom logs had been resting on the ground for so long they were completely rotten. A thousand acres of valley and mountain, a meadow and a creek, and no near neighbours, gave her the living room she needed, but she had to have a livable house and she couldn't live in any but a log house. The problem was how to get a log house built in 1949?

By fortunate chance an old Cariboo cowpuncher happened along. Charlie Morrow had ridden for a number of the big cattle ranches in the interior and had attained the status of head rider at Chilco Ranch in the late nineteen-twenties, but he was now "getting on" and had given up the active life. Mrs. Adler's proposition suited him fine. He would build her log house, taking his own time while savouring the peaceful quiet of the valley. He felled and barked the long fir logs and let them season for a year — then, with one horse to skid and roll them up, he began to build.

Mrs. Adler knew exactly the house she wanted — it was to be simple, honest and rustic. No fancy touches, no slick panelling for interior finish, no frame partitions or modern ceiling tile. Everything was to be of logs. She wanted to look on the bare logs and touch them, inside as well as outside her home. The only concession she made to the modern trend was the two large windows that let in

Above, fireplace in Mrs. Adler's home at Six-mile Creek. 1963. Opposite, a house of hewn logs in the Salmon River Valley, near Falkland. 1956.

Broadaxe and froe.

the southern view of pine trees and mountains. Charlie was pleased with her choice. It was just the kind of house he knew how to build, and had built in the Chilcotin country, and far back in the Coast mountains where I helped him once, building cabins for miners. But when I first met Mrs. Adler the house was built and Charlie had drifted on.

It was a real log house, 41 feet by 32, with log partitions, the cracks filled with small split poles. The ceiling throughout was of split logs laid round side down over peeled poles that ran the length of the building. The window frames and door frames were of split logs, and a Dutch door made of slabs divided the kitchen from the main room, 25 x 15, which contained a stone fireplace built by a rancher friend, Bill Palmer, who rented her mountain range to spring pasture his 2000 sheep. A huge old-fashioned range occupied one end of the kitchen and on the wall above this centre of warmth several shelves accommodated the eighteen cats of the establishment. A wood heater stood in a central hallway which led to several bedrooms on the north side. In the big room a large ornamental coal-oil lamp hung from a roof log, for Mrs. Adler was far beyond the reach of Hydro. There was a full bookcase, and pictures, many pictures; photographs of horsemen and horses, cowboys and ranch people in the fashions of the early years of the century. Around the window frames some ancient geraniums rambled like ivy, adequate proof that winter's breath never penetrated the sturdy walls.

Opposite. Interior of log house of the late Gladys Adler. Little Kingdom Ranch, Six-mile Creek, west side of Okanagan Lake. 1963.

A far cry from Mrs. Adler's comparatively recent log house is the home of Mr. and Mrs. Thomas Johnson whose ranch covers part of a high bench of the Fraser Canyon 28 miles above Lillooet. Known as the China Ranch owing to it being occupied originally by Chinese in gold-rush days, it was, until recently, accessible only by long and tortuous trails. The house is said to be well over a century old, and actually looks older, for its ancient logs, weathered almost black, give an appearance of antiquity far exceeding that of such authentic relics as Cottonwood House. Perhaps a century of the canyon wind was sufficient to produce the look of wrinkled age, for the house stands in the open on high ground, exposed to the weather on every side. The main and original part of the house is composed of large logs, thirteen high — for it is of two stories — very roughly hewn with the axe, but with close-fitting dovetailed corners. The gables are built up with logs and five stout poles carry the roof in fashion generally associated with cabins. A lean-to kitchen with logs left in the round was apparently added at a later date. Since the children grew up and went away the Johnsons no longer use the upper floor of the old house, and the lower rooms allow them ample space for comfortable living. The wood-burning range in the kitchen is the warm heart of a log house, and this old home at China Ranch will never suffer a cold transplant. Hydro is impossibly far away.

Ancient logs, roughly hewn with axe in century old China Ranch house in Fraser Canyon above Lillooet. 1965.

The same can be said for the log house of Mrs. Leona Weeden at McGillivray Creek, high on the mountain above Anderson Lake. It's a typical hermit's cabin. "Gone to your place," read the note tacked to the door the first time I called, a sure sign that another prospector lived in the vicinity. The house is of large logs with the bark left on, saddle notch corners and a roof of cedar shakes, typical of many buildings put up for temporary occupation. A placer-mining company worked the ground in the nineteen-twenties and Mrs. Weeden took over the leases in 1933. "Before that," she told me, "I was trapping in the Kitsumcalum. But mining's the thing. I've prospected all these mountains — back packed and camped all over the place. Be eighty next birthday — fit as a fiddle too. I work these leases and got an interest in a hard rock proposition across the lake — the Queen Anne — silver, lead, zinc. Worked on her last summer. I can sharpen steel, handle powder, drill rock — you can't beat this game, Son." Nice old lady, I thought; the last person I interviewed called me Dad.

Mrs. Weeden lives in her log house now during the summer but winters away down below on the lake shore with relatives. She works her placer lease with water from the creek, flumed to various sections of old creek bed where she finds pay dirt, and her prospector neighbour — he's only seventy — helps her shift the large boulders she encounters.

Painted Chilcotin log cabin with dirt roof near Alexis Creek. 1960.

South front of the Giser home at Celista built about 1909. 1973 photo.

Opposite. Prospector Mrs. Weeden and her log cabin home at McGillivray Creek. 1969.

Shoreholme. The log house of the late H.W. Herridge on Upper Arrow Lake near Nakusp. 1971.

In the Arrow Lakes country six miles north of Nakusp sits the handsome log house of the Herridge family, Shoreholme, well known to erstwhile travellers on the Arrow Lakes. The house was built near the water's edge and was an isolated port of call for the old stern-wheel steamboat Minto on her regular ramblings up and down the lakes. Later, when the Minto was beached, the motor-ship Lardeau took over the ferry service on Upper Arrow Lake, continuing the calls at Shoreholme where the passengers had the opportunity to admire the house in its peaceful setting — a little bay with a sandy beach and a background of wooded mountain.

H.W. "Bert" Herridge was very proud of his log house, and when he became member of parliament for Kootenay West was rather fond of boasting that he was the only Canadian M.P. that lived in a log cabin. He was also in the habit of appending "Log Cabin Socialist" to his signature

on some of his correspondence. But Shoreholme is more than a cabin. It is a roomy house of two stories built with the straight and even poles for which the Arrow Lakes country is noted. The second floor is supported on hewn beams which are an ornament to the large and comfortable living room, a beautiful, friendly room, with a big stone fireplace and walls lined with books.

Like many country homes in the interior, Shoreholme evolved over a period of years to its present state. The logs for the original part of the building were cut in 1936 and construction started the following year. Neighbours and friends had a hand in the building, and Mrs. Herridge and her four daughters split all the cedar shakes for the roof. A kitchen wing and basement were added in later years, in fact, improvements were carried out until 1950 when the house was at last considered to be complete. However, it was not destined to remain on its original site, the lovely waterside location so much admired by the ferry passengers. Threatened by the impending flooding behind Arrow Dam, the log house of Shoreholme was moved by Hydro crews in 1967 to a new location 200 yards away and well above the projected new water level. Set on a new foundation it rests in harmony with its forest background, but its idyllic lake-shore setting can never be restored. The lake is now a reservoir, with periods of draw-down exposing large areas of desolate mud-flats, but fortunately, the house is now far enough from this distressing sight as to retain much of its original charm as a wilderness home.

Above, left. Rear of Giser house. Celista. 1973.

Above, right. Log ranch house and store, previously a roadhouse, at Jesmond. 1962.

Below. Ranch house. Meldrum Creek, west of Williams Lake. 1972.

Opposite, top. Two-storey log house in the upper Columbia Valley south of Golden. 1960.

Opposite, bottom. Mr. and Mrs. Lake in front of their home at Johnson's Landing near Argenta on Kootenay Lake. 1965.

Right, George Elmes' house at Anglemont on north shore of Shuswap Lake, showing detail of dormer windows.

Below, home in mountainside clearing near near Falkland. 1962.

MAINLY BARNS

Log work of some kind is bound to capture the attention of the traveller in the rural areas of interior British Columbia. Here he finds remains of log buildings, log bridges and cribbing on road side and river bank; forest and open country; mountain side and valley. But not all are the wasting relics of early settlement. Many of those in use and occupied today are likely to be of a more recent date. There are still homes, stores and even post offices and churches, blessed with the rustic simplicity of logs. More recently, schools have been 'liberated' from such a 'primitive' state by the ubiquitous trailer.

In the remote districts, one is more likely to come upon a log post office, store or church still occupied and open for business. Happily, the church seems more susceptible to sentimental considerations. The post office seems to be rushing headlong into the computer age.

Overpass for logging operations at Taft, Eagle Valley, now removed. 1962.

Log barns are holding their own splendidly. Modern roofing techniques have extended the life of the barns and hydro power has made them more convenient to work in. Log barns are still pretty common on farms and ranches, but their importance — to the rancher especially — has declined with the mechanical revolution. The big barns — not all of them log — were made to stable many teams of draught horses and accommodate oat bins and a harness room. The lofts were big enough to hold huge quantities of loose hay. In those days, the barn would be the chief centre of activity in the ranchstead. Teamsters had to get out and feed and harness their horses before breakfast. The barn boss did double duty as general custodian and guardian of the oat bin. He was usually an old timer, with whiskers and a lifetime's experience with horses.

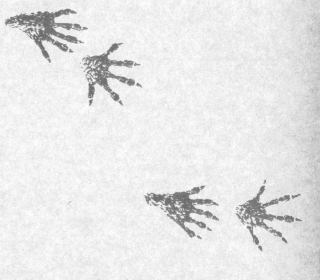

Opposite. Log hay barn near Armstrong. 1950.

Lake Windermere Station at Athelmer, upper Columbia Valley. 1960.

Mountain farm near Vernon. 1950.

Summer found these ranch barns fully tenanted with teams harnessed and ready to start work at seven o'clock, for besides the all-important activities of haying time there was always plenty of work for teams and teamsters. Even ranches had a certain amount of agricultural work to get through in preparing and seeding new hayfields and sometimes growing crops of oats and barley, and there were always hauling jobs. One job peculiar to irrigated ranches was "marking", an operation carried out on the hay and grain fields to distribute the water evenly in the days of furrow irrigation. The implement known as the marker was hauled by a team of horses, and it produced a couple of small compact furrows about three feet apart, the operation being continued until the entire field was engraved with a pattern of parallel rills. Haying time, of course, found every stall in the barn occupied. At that time there would be several teams employed steadily on the mowers, one raking, perhaps half a dozen on the sloops and wagons hauling hay to the stacks, and the inevitable derrick team probably handled by a boy, which worked the creaking apparatus that swung the sling-loads of hay on to the growing stacks.

In winter the barns were warm dens where the teams chomped and stomped when they were not hauling out sleigh loads of hay from the stacks they had helped to build in the summer. Teamsters would be away to the stacks before daylight having harnessed their teams by the dim light of lanterns. Feeding cattle was easier on the horses than the teamsters who had to work hard to fork loads of hay on to the rack while the team stood idle. Later the plodding team would slowly circle the feeding ground followed by a herd of hungry cattle while

Fraser Canyon-style barn with haystack roof north of Lillooet. 1966.

Barn with haystack roof, Big Bar Creek. 1966.

the men spread the hay on the snow. By noon they would be back in the shelter of the barn with perhaps some light afternoon chores like hauling wood ahead of them to finish out the day.

The Cariboo was a great country for horses in the early years of the century. Alkali Lake Ranch bred Clydesdales and had about a hundred; Carson's Ranch on the plateau top of Pavilion Mountain had stables accommodating sixty horses; Chilco Ranch also bred Clydesdales and had 225 horses on the ranch; Dog Creek Ranch had 200 horses from Morgan and Percheron stock. The Grange, near Marble Canyon between Lillooet and Cache Creek, kept 30 to 40 Clydesdales, and Hat Creek Ranch which was occupied with horse breeding on a large scale had stabling for 200 head and bred Morgans and Clydes. 111 Mile Ranch had a hundred head, mostly Clydesdales.

Now the stables are empty and the big barns serve chiefly as housing for baled hay and perhaps shelter for tractors and machinery.

But a simpler type of building, in fact the most primitive of the log barns, still fulfills its original purpose in the drier areas where ranching in a small way occupies the rare and scattered inhabitants. The sinuous tracks scratched over the corrugated landscape of the Fraser Canyon in that wild stretch between Lillooet and Soda Creek lead to hidden hollows where clusters of log buildings may contain a specimen of the roofless barn. This interesting structure could be, and probably often is, mistaken for an incomplete building, for its roofless state certainly fosters that impression. The builders, however, knew what they were about. After erecting four stout log walls, instead of a roof they laid a platform of poles, on which they then built a substantial haystack. It is a neat combination of utility and economy, for the haystack serves the purpose of a roof as well as being the most convenient source of fodder for the animals below. Further, the hay is thus stored well out of reach of hungry "critters" which are notorious for breaking through stackyard fences to hay stored in the fields.

Church at Edgewood. 1958.

Opposite. Log fence and barn. Douglas Lake Ranch. 1972.

Barn near Telkwa. 1964.

Opposite, top. Barn near Aiyansh, Nass River, now converted to Tea House. 1964.

Bottom. Abandoned school, Big Bar Creek, Cariboo. Subject of the book "Buckskin and Blackboard", by Phyllis Taylor, an English girl who came to the isolated spot to teach.

This primitive type of barn — never a very large building — is generally used for stabling a few horses, chiefly for the convenience of having them handy, for ranch stock normally winters in the open. It is known on the Chilcotin jackpine plateau, on small backwoods ranches and on the distant wild hay meadows of the bigger concerns, where a sleigh team and a couple of saddle horses needed accommodation during a winter feeding period. A slight advance on this half-building is the dirt-roofed barn which is common through the Cariboo and Chilcotin. Because it has a roof, albeit of low pitch and supporting a crop of weeds, it has a more finished look, but it lacks storage space for fodder. It is a good place for hanging harness, saddles and gear, but the horses that wear these trappings probably spend more time in the adjoining corral than in the barn itself.

Such low buildings with no lofts would not suit the farmer who depends on substantial barns for sheltering his dairy cattle in winter and milking them at all seasons. For these purposes the log barn as a dairy unit has pretty well had its say, except with some small producers who are not engaged in the production of fluid milk. The log barn is a comfortable place for cattle as the log house is for the husbandman, but the stringent regulations now

Top, small barn at Anglemont. 1973.

Bottom. Abandoned barn in the drybelt near Barnhartvale. 1960.

Opposite. Scene in the Pinantan Lake country, South Thompson Valley. 1971.

governing the production of high quality milk render it unlikely that such a building can be brought up to the required standards.

When settlers were chopping out their clearings and building log homes and barns there were fewer regulations affecting the sale of milk. Farmers could supply milk to the little interior towns direct from their farms so long as they met the legal butterfat standards, and there were no barn inspectors to rule on such matters as light, space and sanitary conditions. The builders of the barns arranged those items to suit their own requirements which were certainly not so demanding as the modern dairyman's. But the interior of a log barn in winter was a cosy place, warmed by the body heat of rows of satisfied cows that flopped down with sighs of repletion after the regular feed of hay. The comforting sound of the champing of horses continued as long as there was fodder in reach of the team. The farm cats began to assemble as milking time drew near, anticipating their allowance of warm milk. When the farmer entered the barn there would be a general uprising of cows accompanied by rattling of chains or stanchions. Bulging eyes would follow him as he climbed the ladder and disappeared into the loft, and heads would toss as hay commenced falling through the hatch. The stalled animals were always ready for their feed despite the inactive life in the barn. There would be a twice-daily emergence for water and exercise though, and a sunning or two in milder weather, but the bulk of a dairy cow's winter was spent under cover, protected from the rigours of the season by the close-laid logs of the barn and the hayloft above.

The whine of the cream separator was then heard in every farmstead, for in those days most interior farmers produced cream for buttermaking, but with increasing population came a demand for fluid milk and a general upgrading of production methods resulted. Log barns could be modernised to a certain degree, and some farmers introduced improvements in the way of steel stanchions, drinking bowls, concrete floors and milking machines, but the growth of the dairy business outpaced them. The present-day dairy plant is a far cry from the barns the settlers built.

Some farmers' log barns still function in the manner intended by their builders although the methods of handling crops have changed. The old-timers filled their large lofts by means of cable equipment worked by horses which hoisted loose hay by the sling-load to the peak of the building, but the modern farmer sends his bales up aloft by means of a light elevator worked by a small motor; and the old cable equipment, which was a great labour saver when it was introduced, now remains unused among the rafters.